STUCK IN SPACE

The Account of NASA's Starliner Mission

Jason T. Morgan

Copyright © 2024 by Jason T. Morgan

All rights reserved. No part of this publication may be reproduced, distributed, or transmitted in any form or by any means, including photocopying, recording, or other electronic or mechanical methods, without the prior written permission of the publisher, except in the case of brief quotations embodied in critical reviews and certain other noncommercial uses permitted by copyright law.

Table Of Contents

Introduction   7
     The Importance of the ISS and Human Spaceflight   8
     Introduction to the Crew: Sunita Williams and
     Butch Wilmore   10

Chapter 1: The Mission Plan   14
     Objectives of the Mission   14
     The Role of the Boeing Starliner   16
     Pre-launch Preparations   18

Chapter 2: The Launch   22
     Countdown and Lift-Off   22
     Initial Successes   23
     The Journey to the International Space Station   25

Chapter 3: Life on the ISS   29
     Daily Routines of Astronauts   29
     Scientific Experiments Conducted   30
     Interactions with Ground Control   32

Chapter 4: The Unexpected Crisis   34
     Discovery of the Starliner Thruster Issues   34
     Initial Reactions from NASA and Boeing   35
     Analysis of the Situation   37

Chapter 5: The Engineering Challenge — 40
Detailed Examination of the Starliner's Propulsion System — 40
Ground Tests vs. In-Orbit Performance — 42
The Struggles of Spacecraft Troubleshooting — 43

Chapter 6: Contingency Plans — 46
Considering the Uncrewed Return of Starliner — 46
Potential Use of SpaceX Crew Dragon for Rescue — 48
Logistics and Challenges of an Extended Mission — 49

Chapter 7: Psychological and Physical Challenges — 52
Impact on Astronauts' Mental Health — 52
Maintaining Morale in Isolation — 53
Managing Supplies and Space Station Resources — 55

Chapter 8: Historical Parallels — 57
Comparison with Previous Space Mission Crises — 57
The Story of Sergei Krikalev on Mir — 59
Lessons Learned from Past Space Incidents — 60

Chapter 9: The Decision — 63
NASA's Call on Starliner's Fate — 63
Preparing for a Potential Extended Stay — 64
The Astronauts' Response and Adaptation — 66

Chapter 10: Public and Media Reaction — 68
How the News Broke — 68

Public Concerns and Speculations	69
Media Coverage and Analysis	70

Chapter 11: The Return — **72**
NASA's Decision-Making Process	72
Preparing for the Return	73
Execution of the Return Plan	74

Chapter 12: Aftermath and Reflections — **76**
Debriefing and Evaluation	76
Impact on Future Space Missions	77
Boeing's Response and Starliner's Future	78

Conclusion — **80**

Introduction

The mission involving NASA astronauts Sunita Williams and Butch Wilmore began as a routine expedition to the International Space Station (ISS) aboard Boeing's new spacecraft, the Starliner. Launched on June 5, 2024, the mission was originally slated to last just eight days, primarily serving as a critical test of the Starliner's capabilities for future manned missions. The objective was to validate the spacecraft's performance, safety, and reliability, thus paving the way for its regular use in transporting astronauts to and from the ISS. However, what was meant to be a straightforward mission quickly turned into an extended ordeal, as unforeseen technical issues with the spacecraft's propulsion system cast doubt on its ability to safely return the crew to Earth.

The Starliner mission was part of NASA's broader strategy to diversify its options for crewed spaceflight, reducing reliance on the Russian Soyuz and complementing the capabilities provided by SpaceX's Crew Dragon. Boeing's Starliner was envisioned as a key component in NASA's Commercial Crew Program, which sought to foster innovation and commercial partnerships to maintain and expand the United States' leadership in space exploration. The mission was also significant in demonstrating the versatility and resilience of the ISS as a platform not just for scientific research, but also for managing and overcoming unexpected challenges in human spaceflight.

The Importance of the ISS and Human Spaceflight

The ISS represents one of humanity's most significant achievements in space exploration, serving as a symbol of international cooperation and a hub for scientific research that cannot be conducted on Earth. Orbiting approximately 400 kilometers above our planet, the ISS has been continuously inhabited since November 2000, with astronauts from around the world contributing to its operation and research endeavors. The station's unique environment allows scientists to conduct experiments in microgravity, which have led to breakthroughs in medicine, materials science, and our understanding of fundamental physics.

Human spaceflight, epitomized by missions to the ISS, is crucial for several reasons. It advances scientific knowledge by providing data that can only be gathered in space, such as the effects of long-term spaceflight on the human body. It also serves as a stepping stone for future exploration missions to the Moon, Mars, and beyond. The ability to live and work in space is essential for these long-duration missions, and the ISS offers a platform to develop and test the necessary technologies and life support systems.

Furthermore, the ISS plays a pivotal role in fostering international collaboration. It is a joint project of NASA, Roscosmos (Russia), ESA (Europe), JAXA (Japan), and CSA (Canada), among others. This collaboration has helped maintain peaceful relations between participating nations and has led to shared knowledge and resources that benefit all of humanity. The ISS is also an inspirational symbol, showing what can be achieved when nations come together for a common goal, and it continues to inspire new generations of scientists, engineers, and explorers.

Introduction to the Crew: Sunita Williams and Butch Wilmore

Sunita Williams and Butch Wilmore, the two astronauts on this mission, are among NASA's most experienced and respected figures. Both have extensive backgrounds in spaceflight, bringing a wealth of knowledge, skills, and resilience to the mission. Their experience would prove crucial as the mission unfolded in unexpected and challenging ways.

Sunita Williams is a veteran astronaut who has spent over 322 days in space, across multiple missions. Born in Euclid, Ohio, on 19th September, 1965, Williams graduated from the United States Naval Academy in 1987 and later earned a master's degree in engineering management. She became a Navy test pilot before being selected as an astronaut by NASA in 1998. Williams has set records for the most spacewalks by a woman and the most time spent on spacewalks, accumulating more than 50 hours outside the ISS. Her previous missions include Expedition 14/15 in 2006 and Expedition 32/33 in 2012, during which she served as a flight engineer and commander, respectively. Williams is known for her calm demeanor, technical expertise, and ability to handle high-pressure situations—a skill set that would be invaluable during the current mission.

Butch Wilmore, born in Tennessee on 29th December, 1962, is another seasoned astronaut with a background in naval aviation. Like Williams,

Wilmore is a graduate of the United States Naval Academy and holds a master's degree in aviation systems. He was selected as an astronaut in 2000 and has flown on two space missions prior to the Starliner mission: STS-129 aboard Space Shuttle Atlantis in 2009 and Expedition 41/42 on the ISS in 2014-2015. During his time on the ISS, Wilmore performed multiple spacewalks and was known for his leadership skills and technical acumen. His previous experience on the ISS, coupled with his engineering background, made him a perfect candidate for this mission, especially given the unexpected challenges that arose.

Together, Williams and Wilmore form a formidable team, bringing a blend of experience, technical expertise, and a deep understanding of space operations to their mission. Their professionalism and resilience in the face of adversity are testaments to the rigorous training and preparation that all astronauts undergo, and they remain a beacon of human capability and determination as they navigate the complexities of space travel.

Chapter 1: The Mission Plan

Objectives of the Mission

The mission that would eventually become a test of endurance and ingenuity for NASA astronauts Sunita Williams and Butch Wilmore began with a straightforward set of objectives. The mission was initially conceived as a key step in NASA's ongoing efforts to expand its crewed spaceflight capabilities, specifically through the use of commercially developed spacecraft. At the heart of this mission was the Boeing CST-100 Starliner, a spacecraft designed to transport astronauts to and from the International Space Station (ISS) as part of NASA's Commercial Crew Program.

The primary objective of the mission was to conduct a comprehensive test of the Starliner's systems in a live operational environment. Although the spacecraft had undergone rigorous testing on the ground, this mission represented its first major trial in space with a crew on board. The mission aimed to validate the spacecraft's performance across several critical phases: launch, docking with the ISS, orbital operations, and safe return to Earth. Each of these phases was crucial in ensuring that the Starliner could reliably serve as a transport vehicle for future crewed missions.

A secondary, yet equally important, objective was to further NASA's research on human adaptation to space. While the mission was initially scheduled for just eight days, any time spent aboard the ISS provided valuable data on how astronauts adjust to microgravity, the effects of long-duration spaceflight on the human body, and the operational challenges that come with living and working in space. This data would be vital for NASA's long-term goals of returning to the Moon under the Artemis program and eventually sending astronauts to Mars.

Finally, the mission aimed to foster closer collaboration between NASA and its commercial partners. The success of the Starliner was not just a matter of pride for Boeing; it was a critical component of NASA's strategy to reduce its reliance on the Russian Soyuz spacecraft and to diversify its crewed spaceflight options. A successful mission would solidify Boeing's role as a key player in the next generation of space exploration and bolster the United States' leadership in space.

The Role of the Boeing Starliner

The Boeing CST-100 Starliner was designed to be a state-of-the-art spacecraft, capable of carrying up to seven astronauts to low Earth orbit (LEO) destinations like the ISS. Unlike NASA's previous spacecraft, the Starliner was developed as part of a public-private partnership, with Boeing investing its own resources alongside NASA's funding. This partnership aimed to leverage the innovation and efficiency of the private sector to meet the space agency's ambitious goals.

The Starliner was equipped with several advanced features designed to enhance crew safety and mission success. One of its key features was the autonomous flight capability, allowing the spacecraft to operate with minimal human intervention. This capability was essential for reducing the workload on astronauts and ensuring that the spacecraft could safely complete its mission even in the event of a communication failure with ground control.

Another critical aspect of the Starliner's design was its emphasis on crew safety. The spacecraft featured an innovative launch abort system that could quickly propel the capsule away from the rocket in case of an emergency during launch. This system was designed to operate at all stages of the launch, ensuring that the crew could escape safely even if a problem occurred late in the ascent. Additionally, the Starliner's robust heat shield and parachute systems were designed to ensure a safe re-entry and landing, whether on land or water.

The Starliner was also intended to be a reusable spacecraft, with each vehicle designed to fly up to ten missions. This reusability was a critical factor in making space travel more economical and sustainable, aligning with NASA's long-term vision of making space more accessible. The spacecraft's modular design allowed for easy refurbishment between missions, ensuring that it could be quickly turned around for subsequent flights.

For this particular mission, the Starliner's role was to serve as both a transport vehicle and a testbed for its various systems. The successful completion of the mission would demonstrate the spacecraft's readiness for regular operational use, paving the way for future missions involving both NASA astronauts and potentially private astronauts in the growing space tourism market.

Pre-launch Preparations

The preparations for the Starliner mission were extensive, reflecting the high stakes involved in testing a new spacecraft with a crew on board. Months before the scheduled launch, a rigorous series of tests and simulations were conducted to ensure that the spacecraft and its systems were ready for the challenges of space.

One of the most critical aspects of pre-launch preparations involved extensive testing of the Starliner's propulsion and navigation systems. Engineers conducted multiple ground-based simulations to replicate the conditions of space as closely as possible, testing the spacecraft's ability to maneuver in orbit, dock with the ISS, and re-enter Earth's atmosphere. These tests were designed to identify any potential issues that could arise during the mission, allowing engineers to make necessary adjustments before launch.

The crew, too, underwent extensive training in preparation for the mission. Williams and Wilmore participated in numerous simulations that mimicked every phase of the mission, from launch to docking with the ISS to emergency procedures. This training was crucial in ensuring that the crew could respond effectively to any situation that might arise, particularly in light of the spacecraft's new and relatively untested systems.

In addition to technical preparations, the mission team also focused on ensuring that all logistical aspects were in place. This included coordinating with the ISS crew, planning for resupply missions, and ensuring that all necessary equipment and supplies were loaded onto the spacecraft. Every detail was scrutinized to minimize the risk of unforeseen complications.

As the launch date approached, final checks were conducted on the spacecraft, the rocket, and the launch facilities. Engineers and mission managers worked around the clock to ensure that everything was in perfect condition for the launch. On the day of the launch, the Starliner stood ready on the launch pad, a testament to the months of hard work and meticulous planning that had gone into the mission.

Despite the extensive preparations, however, no space mission is without risks. The nature of space exploration means that even the most carefully planned missions can encounter unexpected challenges. The Starliner mission was no exception, and the issues that arose during the mission would test the resilience and adaptability of both the spacecraft and its crew.

Chapter 2: The Launch

Countdown and Lift-Off

The Boeing CST-100 Starliner's launch marked a significant milestone for NASA, Boeing, and the future of space travel. The Kennedy Space Center buzzed with anticipation as the launch countdown commenced. Engineers and technicians meticulously ensured that every aspect of the mission was ready. The Starliner sat atop a reliable United Launch Alliance Atlas V rocket, ready to propel it into space.

Amidst the final countdown minutes, Sunita Williams and Butch Wilmore were strapped in the Starliner, fully prepared after years of training. They methodically went through their pre-launch checklists while staying in touch with mission control. The rocket was ready, the launch tower cleared, and all systems were on standby as the liftoff time approached.

The tension mounted as the countdown neared its end. Mission control checked off the final seconds: "Ten, nine, eight, seven, six, five, four, three, two, one... liftoff!" The Atlas V's engines ignited, and the rocket soared into the sky, marking the start of the Starliner's journey.

Initial Successes

The initial phase of the launch proceeded smoothly, with the rocket performing flawlessly as it climbed through the atmosphere. The Atlas V delivered its massive thrust, pushing the Starliner higher and higher into space. As the rocket ascended, it passed through the different stages of flight—Max Q, the point of maximum aerodynamic pressure, was successfully cleared without any issues. The separation of the first stage of the rocket occurred as planned, with the second stage taking over to continue the journey into orbit.

Inside the Starliner capsule, Williams and Wilmore closely monitored the spacecraft's systems. The displays in front of them showed all the critical information they needed: velocity, altitude, and system status. As the rocket ascended, the crew experienced the intense G-forces typical of launch, but they remained focused, their training kicking in as they performed their assigned tasks.

The rocket continued to perform well, and the Starliner capsule reached the edge of space. At this point, the second stage of the Atlas V rocket completed its burn, placing the Starliner into a preliminary orbit. The spacecraft was now in space, and the first major milestone of the mission had been achieved. The Starliner was in a stable orbit, and all systems appeared to be functioning normally.

As the spacecraft separated from the Atlas V rocket, it began its autonomous operations. The Starliner's onboard systems were designed to take over control at this point, using a combination of GPS, star trackers, and other sensors to navigate through space. The spacecraft deployed its solar arrays to generate power, and it adjusted its orientation to prepare for the journey to the International Space Station.

The Journey to the International Space Station

With the successful launch behind them, the focus now shifted to the journey to the International Space Station (ISS). The Starliner was in orbit, but it needed to perform a series of precise maneuvers to align its trajectory with that of the ISS, which was orbiting Earth at an altitude of approximately 400 kilometers.

The first step in this process was the initiation of orbital insertion burns. These burns were small, controlled firings of the Starliner's thrusters designed to adjust the spacecraft's orbit and gradually bring it closer to the ISS. The burns were executed as planned, with each one bringing the

Starliner closer to its rendezvous point. This phase of the mission required utmost precision, as even a small error in timing or thrust could result in a missed approach or a need for corrective maneuvers.

During this phase, the crew continued to monitor the spacecraft's systems, ready to intervene if necessary. The Starliner's autonomous systems were performing well, but Williams and Wilmore remained vigilant, understanding the complexities and potential risks involved in orbital rendezvous. Communication with mission control was constant, with both ground teams and the crew exchanging updates on the spacecraft's status and upcoming maneuvers.

As the Starliner drew closer to the ISS, it entered what is known as the "approach corridor," a precise path that would guide it to the docking port on the space station. The spacecraft's sensors and navigation systems worked in tandem to ensure that it stayed on course. The final phase of the approach required extreme accuracy, as the Starliner needed to slow down and align itself perfectly with the ISS's docking port.

The journey to the ISS was progressing as planned, and everything seemed to be on track for a successful docking. The crew, mission control, and all involved were optimistic, with the initial successes of the mission bolstering their confidence. The Starliner was proving its capabilities in real-time, demonstrating that it could navigate the complexities of space travel and approach the ISS for docking.

However, space missions are never without challenges, and as the Starliner approached the final phase of its journey, the crew and mission control would soon face unexpected difficulties that would test their resilience and problem-solving abilities. The journey to the ISS was not yet over, and the mission's ultimate success would hinge on the ability to overcome the obstacles that lay ahead.

Chapter 3: Life on the ISS

Daily Routines of Astronauts

Life aboard the International Space Station (ISS) is a unique blend of structured routines and the unpredictable nature of space. For NASA astronauts Sunita Williams and Butch Wilmore, their extended stay on the ISS due to the Boeing Starliner's technical issues meant adapting to the station's daily rhythms. Despite the challenges that led to their prolonged mission, the crew maintained a strict daily schedule, which was essential for their physical and mental well-being in the microgravity environment.

Each day began with a wake-up call, followed by personal hygiene routines and breakfast. Even in space, maintaining personal hygiene is critical, with astronauts using rinseless body wipes and waterless shampoo. After breakfast, the crew participated in the first of two daily exercise sessions. Exercise is vital in space, where the absence of gravity leads to muscle atrophy and bone density loss. Williams and Wilmore, like their colleagues, spent about two hours each day on the ISS's treadmill, stationary bicycle, or resistance exercise devices, ensuring their bodies remained in peak condition.

The astronauts' workday consisted of a series of tasks ranging from scientific experiments to maintenance activities. The station's day was punctuated by scheduled meal times and regular communication with ground control, ensuring the crew remained on track with their duties and well-supported by the team on Earth. Despite the demanding schedule, there was also time allocated for personal activities, such as reading, watching movies, or simply enjoying the breathtaking views of Earth from the station's Cupola module.

Scientific Experiments Conducted

One of the primary objectives of the ISS is to serve as a laboratory for scientific research that cannot be conducted on Earth. During their extended mission, Williams and Wilmore were involved in a variety of experiments across different scientific fields, from biology and physics to materials science and Earth observation.

A significant focus was on biomedical research, particularly in understanding how long-term spaceflight affects the human body. The astronauts participated in studies examining muscle atrophy, bone density loss, and changes in cardiovascular health, providing valuable data for future long-duration missions to the Moon and Mars. They also contributed to research on immune system responses in space, a crucial area of study given the potential for astronauts to encounter unknown pathogens during deep-space missions.

In addition to biomedical experiments, Williams and Wilmore conducted research on the behavior of fluids in microgravity, an area of interest for both space exploration and industrial applications on

Earth. Another notable experiment was the testing of advanced materials in space, exposing different substances to the harsh conditions of space to study their durability and potential uses in spacecraft construction.

The ISS also served as a platform for Earth observation. The crew used the station's sophisticated cameras and sensors to monitor environmental changes on Earth, contributing to studies on climate change, natural disasters, and urban development. These observations provided a global perspective that is difficult to achieve from Earth-based platforms.

Interactions with Ground Control

Effective communication with ground control is crucial for the success of any space mission, and for Williams and Wilmore, these interactions were even more significant given the uncertain duration of their stay. The astronauts maintained regular contact with mission control in Houston, engaging in scheduled conferences and real-time problem-solving sessions as needed.

Daily communications involved updates on the crew's health, status reports on scientific experiments, and briefings on upcoming tasks. Ground control also provided emotional support, offering words of encouragement and updates from Earth to keep the crew's morale high during their extended mission.

In the event of technical issues or unexpected challenges, ground control played a critical role in guiding the crew through troubleshooting procedures. For example, when issues arose with the Starliner's thrusters, ground engineers worked tirelessly to analyze the problem, running simulations and providing the astronauts with step-by-step instructions to test and mitigate the issue.

Chapter 4: The Unexpected Crisis

Discovery of the Starliner Thruster Issues

The mission of Sunita Williams and Butch Wilmore aboard the International Space Station (ISS) took an unforeseen turn when engineers detected anomalies with the Boeing Starliner spacecraft's thrusters. The Starliner, which had safely transported the astronauts to the ISS, was expected to bring them back to Earth after their mission. However, during routine checks and post-launch diagnostics, it became apparent that several of the Starliner's thrusters were not functioning as expected.

The thruster issues were initially identified during an automated system check when some of the spacecraft's propulsion units failed to respond.

Further analysis revealed that certain thrusters had powered down unexpectedly, raising concerns about the spacecraft's ability to execute a controlled descent and re-entry into Earth's atmosphere. These thrusters were crucial for the spacecraft's maneuverability, particularly during the deorbit burn and re-entry phases of the return flight.

Initial Reactions from NASA and Boeing

The discovery of the thruster issues sent shockwaves through both NASA and Boeing. For Boeing, the Starliner program was already under intense scrutiny due to previous delays and challenges. The news that the spacecraft might not be fit to bring the crew home safely was a significant setback. Engineers at Boeing immediately began working around the clock to diagnose the root cause of the thruster malfunction. The company issued a statement acknowledging the issue and emphasized its commitment to ensuring the safety of the astronauts.

At NASA, the response was one of cautious concern. The agency's first priority was the safety of Williams and Wilmore. NASA's space operations team quickly convened with Boeing engineers to assess the situation. Senior officials, including Ken Bowersox, NASA's director of space operations, were briefed on the potential risks and the contingency plans that might need to be implemented.

Publicly, NASA assured the media and the public that the astronauts were safe on the ISS and that the situation was under control. However, internally, there was an understanding that this issue could delay the astronauts' return to Earth significantly. The space agency's experience with handling in-orbit crises was brought to bear as they began to explore alternative options for the crew's safe return.

Analysis of the Situation

As the investigation into the thruster issues progressed, it became clear that the problem was more complex than initially thought. The thrusters had been subjected to rigorous testing on the ground, yet these issues only manifested in the space environment. This led engineers to suspect that the unique conditions of space, such as the microgravity environment or temperature extremes, might be influencing the thrusters' performance in ways not fully understood.

Boeing's engineering team initiated a series of diagnostic tests and simulations to replicate the conditions that might have caused the thrusters to fail. Simultaneously, they explored potential fixes that could be implemented remotely or during an uncrewed return mission. There was even consideration given to sending a repair crew to the ISS, though this was viewed as a last resort due to the complexities and risks involved.

Meanwhile, NASA's analysis focused on the implications for the crew's safety and the broader mission objectives. A range of scenarios was evaluated, from attempting a return in the Starliner to extending the crew's stay on the ISS until a safer alternative could be arranged. The idea of sending a SpaceX Crew Dragon spacecraft as a backup return vehicle gained traction, given its proven track record and capacity to safely bring the astronauts home.

Throughout the crisis, both NASA and Boeing maintained open lines of communication, not only with each other but also with the astronauts. Williams and Wilmore were kept informed of the situation, ensuring that they were fully prepared for whatever course of action would be necessary. The collaboration between NASA and Boeing during this unexpected crisis underscored the complexity and inherent risks of space exploration, highlighting the importance of meticulous planning, rigorous testing, and the ability to adapt in the face of unforeseen challenges.

Chapter 5: The Engineering Challenge

Detailed Examination of the Starliner's Propulsion System

The Boeing Starliner's propulsion system, a critical component designed to ensure the safe return of astronauts to Earth, became the focal point of intense scrutiny following the discovery of thruster malfunctions. The Starliner is equipped with multiple sets of thrusters, each playing a vital role in various phases of the mission, from orbital maneuvers to re-entry. These thrusters are designed to provide precise control over the spacecraft's attitude and trajectory, ensuring that it can perform complex operations like docking with the International Space Station (ISS) and safely descending through the Earth's atmosphere.

The propulsion system in question consists of four primary thruster clusters, each containing several individual thrusters. These thrusters rely on hypergolic fuels—chemicals that ignite on contact with each other—allowing for reliable ignition without the need for an ignition system. This design, while robust and proven in other spacecraft, presented unique challenges for the Starliner.

The thrusters had passed all ground-based tests before launch, but once in orbit, a subset began to exhibit failures. This prompted engineers to conduct a thorough examination of the entire propulsion system, including the fuel lines, valves, and the thrusters themselves. The goal was to identify any potential causes of the malfunction that could compromise the safety of the crew during re-entry.

Ground Tests vs. In-Orbit Performance

One of the most perplexing aspects of the situation was the disparity between the thruster performance during ground tests and in-orbit conditions. On Earth, Boeing subjected the thrusters to extensive testing in vacuum chambers designed to simulate space conditions. These tests included thermal cycling, vibration testing, and prolonged operational burns to ensure the thrusters could handle the stresses of space travel. Despite these rigorous tests, the issues only manifested once the spacecraft was in orbit, highlighting the inherent challenges in predicting space-based performance from ground-based simulations.

In-orbit conditions differ significantly from those on Earth. Microgravity, vacuum, and extreme temperature variations are all factors that can affect spacecraft systems in unpredictable ways. The thrusters' exposure to prolonged microgravity might have caused changes in fuel flow or pressure, leading to the malfunctions observed. Additionally, the thermal environment in orbit, where the spacecraft experiences alternating periods of intense heat and cold, could have affected the thrusters' components in ways not fully anticipated during ground tests.

This discrepancy between expected and actual performance underscores the difficulty of replicating the true conditions of space on Earth. While ground tests are essential for initial validation, they cannot fully simulate the complexities of the space environment, making in-orbit testing an equally critical part of spacecraft validation.

The Struggles of Spacecraft Troubleshooting

Troubleshooting spacecraft issues is a formidable challenge, especially when the spacecraft is hundreds of kilometers above Earth. Engineers on the ground must rely on telemetry data sent back from the spacecraft, which can sometimes be limited in detail. In the case of the Starliner, interpreting this data to diagnose the thruster issues required a combination of real-time analysis and simulated recreations of the problem on Earth.

Boeing's engineering team faced the daunting task of determining whether the thruster malfunctions were due to a design flaw, a manufacturing defect, or an unforeseen interaction between the spacecraft's systems and the space environment. This process involved running detailed simulations, conducting additional tests on identical thruster models on the ground, and analyzing every piece of data from the spacecraft.

One of the most significant struggles was the inability to physically inspect the problematic thrusters. Unlike terrestrial systems, where engineers can disassemble and examine faulty components, space systems must be diagnosed and repaired remotely. This limitation required Boeing and NASA to consider various potential solutions, including software patches, adjusting operational parameters, or, in the worst case, planning an uncrewed return of the Starliner.

The engineering challenge presented by the Starliner's thruster issues exemplifies the complexities and risks of space exploration. It highlights the need for continuous innovation in testing and simulation technologies and the importance of flexibility in mission planning to accommodate the unpredictable nature of space travel.

Chapter 6: Contingency Plans

Considering the Uncrewed Return of Starliner

As the thruster issues on the Boeing Starliner spacecraft became apparent, NASA and Boeing were forced to confront a grim possibility: the Starliner might not be safe for the astronauts to use on their return journey to Earth. This scenario required serious consideration of an uncrewed return for the spacecraft, which would entail sending the Starliner back to Earth without Sunita Williams and Butch Wilmore onboard.

The decision to consider an uncrewed return of Starliner was not taken lightly. The spacecraft was designed with multiple redundancies and safety systems to ensure it could safely transport astronauts. However, the unidentified issues with the propulsion system raised concerns about the risk of an uncontrolled re-entry or a failure during the critical descent phase. An uncrewed return would allow engineers to safely bring the Starliner back to Earth for detailed analysis, potentially uncovering the root cause of the thruster malfunctions.

If Starliner were to return autonomously, NASA would need to devise a new plan for returning Williams and Wilmore to Earth. The complexity of orchestrating such a mission was immense, involving not only the technical challenges of ensuring Starliner's safe return but also the logistical and operational implications for the International Space Station (ISS) crew and ongoing mission objectives.

Potential Use of SpaceX Crew Dragon for Rescue

With the Starliner's capabilities in question, NASA turned to another spacecraft in its arsenal: the SpaceX Crew Dragon. Unlike Starliner, the Crew Dragon had already proven its reliability in multiple crewed missions to the ISS. As a result, NASA began evaluating the possibility of using the Crew Dragon for a potential rescue mission to bring Williams and Wilmore back to Earth.

This contingency plan would involve launching a Crew Dragon to the ISS with a minimal crew, likely two astronauts, to maintain mission flexibility. The Crew Dragon would then dock with the ISS, where Williams and Wilmore would board for their return trip. This plan would ensure that the two astronauts could return safely, even if the Starliner was deemed unsafe for crewed flight.

While the Crew Dragon offered a viable alternative, it also presented logistical challenges. The spacecraft's seating capacity and payload limits meant that careful planning would be required to ensure the safe and efficient return of the astronauts, along with any essential cargo. Additionally, coordinating the launch of another spacecraft involved working closely with SpaceX to align schedules and resources, all while managing the ongoing operations of the ISS.

Logistics and Challenges of an Extended Mission

As contingency plans were being developed, one of the most pressing considerations was the impact on the ISS mission if Williams and Wilmore were required to stay longer than planned. Initially scheduled for an eight-day mission, the astronauts were now facing the possibility of an extended stay that could last several months, potentially until February 2025.

An extended mission required careful management of ISS resources, including food, water, and other consumables. Although the ISS is well-stocked, prolonged missions put additional strain on these resources, necessitating resupply missions and careful rationing. The space station's systems, including life support, power, and waste management, were also designed to support a specific crew size, meaning that adjustments would be needed to accommodate the extended presence of Williams and Wilmore.

Another challenge was the psychological and physical well-being of the astronauts. Extended missions in space can lead to increased fatigue, stress, and health risks, particularly in a cramped environment like the ISS. To mitigate these risks, NASA worked to integrate Williams and Wilmore into the daily routines of the station, keeping them engaged in scientific experiments and maintenance tasks to maintain morale and physical fitness.

Moreover, the extended mission meant that Williams and Wilmore would miss important personal milestones and family events, adding emotional stress to the technical and operational challenges. NASA's support systems, including regular communication with family members and psychological support, played a crucial role in helping the astronauts manage these pressures.

Chapter 7: Psychological and Physical Challenges

Impact on Astronauts' Mental Health

Space missions, even under ideal conditions, impose significant psychological stress on astronauts. The sudden extension of Sunita Williams and Butch Wilmore's mission aboard the International Space Station (ISS) due to the Starliner's thruster issues introduced an additional layer of uncertainty and strain. Initially set for an eight-day mission, the prospect of spending several more months in space raised concerns about the potential mental health impacts on the crew.

Prolonged space missions can exacerbate feelings of isolation, confinement, and separation from loved ones. The astronauts, who were mentally prepared for a short-term mission, now had to recalibrate their expectations and adapt to an extended stay.

This shift can lead to increased anxiety, stress, and even depression, as the unpredictability of their return to Earth looms over their daily lives.

NASA has long recognized the importance of mental health in space missions, providing astronauts with resources to manage stress and emotional well-being. Regular communication with family members, access to mental health professionals, and engagement in recreational activities are all part of NASA's strategy to support astronauts' mental health. For Williams and Wilmore, these resources became even more critical as they adjusted to their extended mission.

Maintaining Morale in Isolation

Maintaining morale in the isolated environment of the ISS is essential for the crew's overall well-being and mission success. The sudden extension of their mission required Williams and Wilmore to find ways to stay motivated and engaged, despite the challenges of prolonged isolation.

Routine and structure play vital roles in sustaining morale. The astronauts' daily schedules are meticulously planned, with time allocated for scientific research, station maintenance, exercise, and leisure. Maintaining this routine helped Williams and Wilmore stay focused and productive, even as the uncertainty of their return home persisted.

Additionally, the sense of purpose derived from their work on the ISS contributed to their morale. Engaging in meaningful scientific experiments, contributing to the ongoing success of the mission, and being part of a historic moment in space exploration provided the astronauts with a sense of accomplishment and motivation. Furthermore, the camaraderie among the crew members on the ISS offered emotional support, as they could rely on one another for companionship and understanding.

NASA also encouraged the astronauts to partake in recreational activities to alleviate stress and boost morale. Whether watching movies, listening to music, reading, or simply gazing at Earth from the station's windows, these moments of relaxation provided crucial mental breaks from the demands of the mission.

Managing Supplies and Space Station Resources

With the mission's unexpected extension, managing the ISS's supplies and resources became a top priority. The space station is equipped to support a specific crew size for a predetermined duration, and extending the mission added pressure to ensure that essential supplies like food, water, and medical supplies were sufficient.

NASA's meticulous planning and inventory management systems were put to the test as they assessed the station's resources. Fortunately, a recent resupply mission had delivered additional provisions, but careful rationing and resource allocation were still necessary to avoid shortages. The astronauts had to be mindful of their consumption, ensuring that they conserved resources while still meeting their nutritional and health needs.

The space station's life support systems, including air and water recycling, also had to be closely monitored. Any malfunction or depletion of these critical systems could have severe consequences for the crew's health and safety. Regular maintenance checks and troubleshooting were essential to ensure that these systems functioned optimally throughout the extended mission.

Moreover, the physical constraints of the ISS, with its limited space and shared facilities, required the astronauts to adapt to a more crowded and potentially stressful environment. Effective communication, cooperation, and adherence to protocols were crucial in managing these challenges and ensuring the mission's continued success.

Chapter 8: Historical Parallels

Comparison with Previous Space Mission Crises

Space exploration is fraught with uncertainty, and over the decades, astronauts and cosmonauts have encountered numerous crises in orbit. The situation faced by Sunita Williams and Butch Wilmore, potentially stranded aboard the International Space Station (ISS) due to issues with the Boeing Starliner, bears striking similarities to earlier space mission crises. Each of these incidents underscores the inherent risks of space travel and the extraordinary resilience required of those who venture beyond Earth.

One of the most famous examples of a space mission crisis is the Apollo 13 incident in 1970. What was supposed to be NASA's third crewed mission to the Moon turned into a life-or-death struggle when an oxygen tank explosion severely damaged the spacecraft. The mission's objective quickly shifted from lunar exploration to the safe return of the crew. The Apollo 13 crisis highlighted the importance of adaptability, teamwork, and the ability to troubleshoot under extreme pressure—qualities that Williams and Wilmore, as seasoned astronauts, also had to draw upon during their extended stay on the ISS.

Another comparable incident occurred during the Gemini 8 mission in 1966, when astronauts Neil Armstrong and David Scott experienced a malfunction in their spacecraft's attitude control system. The spacecraft began to spin uncontrollably, but Armstrong's quick thinking and calm response averted disaster. Like the crew of Gemini 8, Williams and Wilmore faced a situation where their training and composure were critical to overcoming unforeseen challenges.

The Story of Sergei Krikalev on Mir

The situation faced by Williams and Wilmore can be likened to that of Russian cosmonaut Sergei Krikalev, who was stranded aboard the Mir space station during the collapse of the Soviet Union in 1991. Krikalev's mission was supposed to be routine, but due to political turmoil, his return was delayed significantly. He ended up spending 311 days on Mir, way longer than the initial five-month plan, setting a record for the longest single spaceflight at the time.

Krikalev's ordeal shares many similarities with the situation faced by Williams and Wilmore. Both missions involved extended stays in space far beyond the original plan, placing immense physical and psychological strain on the astronauts. Krikalev had to deal with deteriorating conditions on Mir, limited communication with Earth, and the uncertainty of when—or even if—he would be able to return home. His experience serves as a powerful reminder of the unpredictability of space missions and the need for astronauts to remain resilient in the face of adversity.

Just as Krikalev had to deal with prolonged isolation and physical strain, Williams and Wilmore also faced similar challenges. Krikalev's narrative emphasizes the need for maintaining purpose and discipline during extended missions, qualities that were essential for Williams and Wilmore as they carried out their duties on the ISS despite the uncertainty surrounding their return.

Lessons Learned from Past Space Incidents

The crises faced by past astronauts have provided valuable lessons for space agencies and future missions. One key lesson is the importance of thorough preparation and training. Astronauts undergo rigorous simulations and emergency drills to prepare for potential crises, allowing them to respond effectively when unexpected situations arise. The experiences of Apollo 13, Gemini 8, and Sergei Krikalev demonstrate that even the most meticulously planned missions can encounter unforeseen challenges, and the ability to adapt quickly is crucial.

Another lesson is the need for robust communication between the spacecraft and mission control. During the Apollo 13 crisis, the collaboration between the astronauts and ground teams was instrumental in developing solutions to the spacecraft's problems. Similarly, ongoing communication with Earth was vital for Krikalev's survival on Mir. For Williams and Wilmore, maintaining strong communication with NASA and other space agencies was essential for managing the Starliner crisis and exploring contingency options.

Finally, these past incidents underscore the importance of psychological resilience in space. The ability to maintain focus, stay calm under pressure, and find motivation even in the face of uncertainty is critical for astronauts on long-duration missions. The challenges faced by Williams and Wilmore on the ISS, much like those encountered by their predecessors, reinforce the need for continued support for the mental well-being of astronauts during extended spaceflights.

Chapter 9: The Decision

NASA's Call on Starliner's Fate

After weeks of intensive analysis, NASA's engineering teams, along with Boeing's experts, concluded that the thruster issues were too significant to ignore. Despite the spacecraft functioning well in some tests, the risk of a malfunction during re-entry was deemed unacceptable. NASA officials had to weigh the immediate safety of the astronauts against the operational and reputational impact of grounding the Starliner.

Ultimately, NASA decided that the safest course of action was to send the Starliner back to Earth without its crew. This decision, while disappointing for the Starliner program, prioritized astronaut safety above all else. NASA's cautious approach reflected its commitment to ensuring that no mission, however significant, would compromise the well-being of its crew members. By opting for an uncrewed return, NASA allowed engineers the opportunity to thoroughly investigate and resolve the issues on the ground, thereby preserving the possibility of future missions with a fully reliable spacecraft.

Preparing for a Potential Extended Stay

With the decision to leave the Starliner in orbit without its crew, the focus shifted to preparing the astronauts, Sunita Williams and Butch Wilmore, for an extended stay on the ISS. Initially planned as an eight-day mission, the astronauts now faced the possibility of remaining in space for several more months, potentially until February 2025. NASA quickly assessed the station's capacity to sustain this extended mission.

The ISS, equipped with sophisticated life support systems, was deemed fully capable of supporting the astronauts for the additional duration. Recent resupply missions had provided ample food, water, and other essential supplies, ensuring that the crew would not face shortages. However, maintaining the physical and mental well-being of the astronauts during this unexpected extension required careful planning.

NASA's mission planners worked closely with the astronauts to adapt their schedules and routines. The extended mission meant that Williams and Wilmore would be integrated more fully into the ongoing scientific research aboard the ISS, while also contributing to the station's maintenance and operations. Their extensive experience in space was invaluable, allowing them to handle the demands of a longer mission with confidence.

The Astronauts' Response and Adaptation

Sunita Williams and Butch Wilmore, both seasoned astronauts, approached the situation with the professionalism and resilience that characterized their careers. While the extension of their mission was unexpected, they quickly adapted to the new circumstances. Understanding the importance of their role in testing and maintaining the ISS's systems, they embraced the challenge with a sense of duty and focus.

Maintaining morale was a critical aspect of their adaptation. The astronauts relied on their training, their strong working relationship, and regular communication with ground control to stay connected with life on Earth. NASA's psychological support teams played an essential role in helping the astronauts manage the emotional and mental challenges of an extended stay in space, ensuring that they remained in good spirits despite the uncertainty.

As the weeks turned into months, Williams and Wilmore continued to contribute to the ISS's operations, conducting scientific experiments, performing maintenance tasks, and supporting their fellow crew members. Their adaptability and positive response to the situation underscored the qualities that make astronauts exceptional: resilience, resourcefulness, and an unwavering commitment to the mission, regardless of the challenges that arise.

Chapter 10: Public and Media Reaction

How the News Broke

Shortly after NASA decided to pause the return of the Starliner spacecraft with the crew on board, reports emerged regarding the thruster issues it was facing. The information was first released by official NASA channels, detailing the problems identified, the precautions taken for the astronauts' safety, and the consideration of an extended stay on the ISS. The statement emphasized NASA's dedication to safety and open communication, noting the possibility of a prolonged mission while reassuring the public that the astronauts were not in immediate jeopardy.

As the news circulated, it quickly caught the attention of major news outlets, space-focused media, and social networking platforms. The seriousness of the situation, combined with public interest in space exploration, ensured a swift spread of the story. Headlines about the "stranded astronauts" and the "Boeing Starliner crisis" dominated media coverage within hours, sparking a global discussion about the event.

Public Concerns and Speculations

The public response included a mix of worry, interest, and speculation when issues arose with the astronauts' spacecraft. Many were concerned about the astronauts' safety and the potential outcomes of the situation. Social media was abuzz with inquiries, worries, and words of encouragement for Williams and Wilmore.

As discussions unfolded, some raised doubts about Boeing's capabilities and the overall safety of the Starliner program. Critics of NASA's partnership with commercial entities took the opportunity to voice their concerns about the privatization of space travel. Speculation also focused on how the incident could affect future missions, such as the planned Artemis missions to the Moon. The unresolved problems with the spacecraft's thrusters led to various theories, some more rooted in imagination than reality.

Despite the apprehension, there was a surge in support and respect for the astronauts. Many expressed confidence in Williams and Wilmore, highlighting their extensive training and expertise. The public's belief in the astronauts' skills and determination emerged as a prevalent theme, serving as a counterbalance to the uncertainties and skepticism seen online.

Media Coverage and Analysis

The media played a crucial role in shaping the public's understanding of the situation, with major news outlets providing constant updates and featuring expert analysis from individuals such as former astronauts, aerospace engineers, and space industry insiders. These experts helped explain the technical aspects of the Starliner's thruster issues, the contingency plans in place, and the potential outcomes that could occur.

Different media outlets had varying tones and perspectives. Some were critical, questioning Boeing's reliability and whether the Starliner was adequately tested before its mission, while others focused on the human element, highlighting the astronauts and their families, and highlighting the risks involved in space exploration.

There were documentaries and detailed reports that explored the incident's broader implications for NASA's commercial partnerships and the future of space travel. Additionally, the situation led to discussions about the challenges of upholding public trust in space missions and the importance of transparency during times of crisis.

Chapter 11: The Return

NASA's Decision-Making Process

The decision-making involved multiple layers of assessment. NASA's primary concern was the safety of the astronauts, which meant evaluating the Starliner's reliability and the feasibility of an uncrewed return. An internal review process, including consultations with engineers, mission planners, and safety experts, was critical in determining whether the Starliner could safely return the astronauts to Earth. Discussions included evaluating if the spacecraft could be safely flown back without crew or if alternative arrangements needed to be made.

In parallel, NASA had to weigh the risks and benefits of utilizing SpaceX's Crew Dragon for a potential rescue mission. This included coordinating with SpaceX to understand the availability and readiness of the Crew Dragon, as well as the implications of adjusting crew assignments on upcoming missions. By mid-August, NASA was expected to make a final call based on the ongoing evaluations and technical assessments, ensuring the decision would prioritize the astronauts' safety and mission objectives.

Preparing for the Return

Preparation for the astronauts' return involved meticulous planning and coordination among various teams. With the possibility of extending their stay, NASA needed to ensure that the astronauts had sufficient supplies and support to sustain them for the additional months. This included logistical arrangements for resupply missions, ensuring that they had adequate food, equipment, and personal items to remain comfortable and effective in their roles on the ISS.

Preparation also involved addressing the spacecraft's technical issues. If the decision was made to return the astronauts aboard an uncrewed

Starliner, NASA had to ensure the spacecraft was configured properly for such a flight. This involved extensive testing and verification to ensure that the Starliner could complete its journey safely without crew onboard. Concurrently, contingency plans for using SpaceX's Crew Dragon needed to be finalized, including modifications to the Crew-9 mission to accommodate the return of Williams and Wilmore.

Training and simulations were also critical components of preparation. The astronauts, as well as the mission control teams, had to be prepared for any scenarios that might arise during the return. This involved updating protocols and rehearsing procedures to handle the complexities of the extended mission and the eventual return, whether it was aboard a modified Starliner or Crew Dragon.

Execution of the Return Plan

The execution of the return plan was a significant milestone in addressing the unexpected challenges faced by the Starliner mission. If the decision was made to return the astronauts with Crew Dragon, a detailed sequence of events was set into motion. This included coordinating with SpaceX for the timely launch of Crew-9, which would need to be adjusted to accommodate the additional seats required for the return journey. NASA would have to ensure that the Crew Dragon mission could integrate smoothly with the current crew rotation on the ISS.

The return itself involved a series of carefully orchestrated steps. The Crew Dragon would launch, dock with the ISS, and transfer Williams and Wilmore back to the spacecraft. Detailed procedures for undocking, re-entry, and landing were established to ensure a safe return to Earth. The astronauts would undergo final checks and preparations before leaving the ISS, with mission control closely monitoring every phase of their journey home.

Chapter 12: Aftermath and Reflections

Debriefing and Evaluation

As the situation with the Starliner spacecraft continues to evolve, NASA is preparing for a comprehensive debriefing and evaluation process once astronauts Suni Williams and Butch Wilmore eventually return to Earth. This debriefing will not only focus on the astronauts' personal experiences and challenges during their extended stay on the ISS but also on the technical performance of the Starliner spacecraft. The detailed assessment will involve analyzing the spacecraft's systems, identifying the root causes of the thruster issues, and understanding how these problems were managed in real-time. The insights gained will be critical in refining spacecraft design, operational procedures, and contingency planning for future missions. NASA's goal will be to extract every possible lesson from this mission to enhance the safety and reliability of human spaceflight.

Impact on Future Space Missions

The extended mission of Williams and Wilmore has already provided NASA with invaluable data on the psychological and physical effects of long-duration spaceflight. Prolonged exposure to the microgravity environment, coupled with the uncertainty surrounding their return, has tested the astronauts in ways that go beyond the planned mission parameters. This experience is likely to influence the planning and execution of future missions, particularly those involving potential mission extensions or unplanned contingencies. The data gathered will contribute to better training and preparation of astronauts, ensuring that they are equipped to handle unexpected challenges. Additionally, the experience has highlighted the importance of robust contingency plans and the need for spacecraft that can perform reliably under extended mission conditions. The knowledge gained will undoubtedly shape NASA's approach to future exploration missions, especially those targeting deeper space and longer durations.

Boeing's Response and Starliner's Future

The problems encountered with the Starliner spacecraft have put Boeing under intense scrutiny. The company's ability to address these issues effectively will determine the future of the Starliner program and its role in NASA's human spaceflight endeavors. Boeing's response will be critical in rebuilding trust with NASA and the broader space community. They will need to not only fix the current issues but also implement long-term solutions to prevent similar problems in future missions. This situation presents a significant challenge but also an opportunity for Boeing to demonstrate its commitment to advancing space exploration and ensuring the safety of the astronauts who depend on their technology. The outcome of this situation will have lasting implications for Boeing's reputation and the future of commercial spaceflight partnerships.

Conclusion

The Boeing Starliner mission is a significant milestone in the ongoing narrative of human space exploration, displaying the challenges and victories that come with venturing into the final frontier. This narrative follows astronauts Suni Williams and Butch Wilmore from their departure to the International Space Station (ISS) through their extended stay and eventual return, delving into the complexities of their mission, unexpected hurdles faced, and the subsequent adaptations made for their safety and success.

This incident underscores the complexities of space exploration, emphasizing the importance of thorough contingency planning due to the unforeseen technical difficulties with the Starliner spacecraft. NASA and Boeing's quick reaction, combined with the astronauts' resilience, showcases the dedication and flexibility required in space missions. It highlights that regardless of meticulous planning and advanced technology, space missions are full of uncertainties that demand adaptability and innovation.

Williams and Wilmore's perseverance during their extended time in space demonstrates the extraordinary resilience of astronauts. Their ability to adjust to changing circumstances, maintain their well-being and determination, and carry on with their critical work on the ISS speaks to humanity's relentless pursuit of exploration and understanding. This narrative not only applauds their accomplishments but also fosters a deep appreciation for the sacrifices and obstacles faced by those who explore beyond Earth.

Looking forward, the insights gained from this mission will undoubtedly influence the future of space travel, informing spacecraft design, mission preparation, and contingency strategies to enhance human spaceflight. As we continue to push the boundaries of exploration, these experiences will guide us in creating safer, more dependable space missions and preparing for the thrilling possibilities that await.

This narrative is a tribute to the spirit of exploration and the continuous pursuit of advancement. The Boeing Starliner mission, with its tests and triumphs, embodies the enduring pursuit of exploring the universe and the unwavering dedication of those who make it possible. Reflecting on this journey reminds us of the vast opportunities beyond our planet and the remarkable achievements humanity can reach through perseverance, innovation, and cooperation. The future of human space exploration holds great potential, and the experiences chronicled here will pave the way for new discoveries and progress in our ongoing exploration of the universe.

www.ingramcontent.com/pod-product-compliance
Lightning Source LLC
Chambersburg PA
CBHW070355230526
45471CB00006B/2576